Over the horizon
A Journey into Infinity
of the Universe and the Mind
Eng. Das Warhe

Eng. Das Warhe, 2024
All rights reserved.
Reproduction is permitted for educational and non-commercial purposes only, provided that the source is cited
April 2024

Table of Contents

Summary 6

Introduction: Exploring the Infinity of the Universe and the Human Mind 13

Chapter 1: Journey into Cosmic Infinity 17

The Galaxies: Universes in Miniature 17

The Milky Way: Our Spiral Galaxy 18

The Stars: Celestial Furnaces of Light and Life 18

The Planets: Worlds to Explore 19

Black Holes: Portals to Infinity 20

Conclusion 20

Chapter 2: Beyond Time and Space 22

The Concept of Time: A Persistent Illusion 22

The Theory of General Relativity: Curvature of Space and Time 23

Quantum Physics: Where Time Hits the Wall of Uncertainty 24

Philosophical Implications: The Nature of Existence and Reality 24

Conclusion 25

Chapter 3: Depth of Consciousness 27

Subjective Experience: The Theater of Consciousness 27

The Philosophy of Consciousness: The Enigma of Being ... 28

Neuroscience of Consciousness: Discovering Neural Circuits..29

Awareness and the Self: The Intertwining of Consciousness 30

Conclusion 31

Chapter 4: Virtual Reality and Mental Illusions..................32

Virtual Reality: Intertwined Dreams and Reality....................33

Optical Illusions: When the Eyes Deceive the Mind............33

Cognitive Bias: When the Mind Deceives Itself 34

The Philosophy of Reality: In Search of the Truth 35

Conclusion 35

Chapter 5: The Search for Extraterrestrial Life 37

The Significance of Space Exploration: The Journey into the Unknown 38

Exoplanets: Potentially Habitable Worlds 38

The Search for Signs of Life: From Chemistry to Biology ...39

The Philosophical Significance of the Search for Extraterrestrial Life: Reflections on Humanity and the Cosmos 40

Conclusion 41

Chapter 6: Mysteries of Existence 42

The Why Question: Searching for Meaning in Human Existence..43

The Inner Journey: Discovering the Deep Self 43

The Reality of Being: Beyond Superficial Appearances.......44

The Question of One and All: The Mystery of the Universe 45

Conclusion 45

Chapter 7: The Creative Mind 47

The Source of Creativity: Birth of the Creative Idea 48

The Psychology of Creativity: The Mental Processes Behind Innovation 48

The Art of Creativity: Expression and Innovation 49

The Philosophy of Creativity: Exploring the Frontiers of the Imagination 50

Conclusion 50

Chapter 8: Knowledge and Ignorance 52

The Challenge of Knowledge: Navigating the Sea of Unknown 52

The Nature of Ignorance: Exploring the Limits of Human Understanding 53

The Illusions of Knowledge: Exploring the Traps of Certainty 54

The Paradox of Enlightened Ignorance: The Awareness of Our Ignorance 54

Conclusion 55

Chapter 9: The Beauty of Art 56

The Expression of the Soul: Art as a Vehicle of Emotions and Thoughts 57

The Wonder of Creativity: Art as an Act of Innovation and Discovery 57

Art as a Reflection of Nature: Exploring the Beauty of the Natural World 58

Beauty as a Portal to Transcendence: Art as a Spiritual Experience 58

Conclusion 59

Chapter 10: The Power of Compassion 61

The Nature of Compassion: The Expression of Universal Love 61

Compassion as a Transforming Force: The Power of Altruism and Solidarity 62

Compassion in Spiritual Practice: The Heart as the Center of Wisdom 62

Compassion as a Path to Wisdom: Learning from the Experience of Others 63

Conclusion 64

Conclusion: The Triumph of Exploration 65

Summary

In the book "Over the Horizon: A Journey into the Infinity of the Universe and Mind", we immerse ourselves in an epic adventure through the infinity of the external universe and the complexity of the human mind. Through a compelling combination of scientific storytelling and philosophical reflections, we will explore the deepest mysteries of existence and venture into unknown worlds in both the cosmos and the human psyche.

Introduction: Exploring the Infinity of the Universe and the Human Mind

In the book's introduction, we delve into the essence of exploring the universe and the human mind. Through a journey of discovery and wonder, we prepare to face the great mysteries that await us, both in the immensity of sidereal space and in the depths of human consciousness. As Carl Sagan says, "We are made of stellar matter. We are a way for the universe to know itself."

Chapter 1: Journey into Cosmic Infinity

In this chapter, we dive into the wonders of the external universe. We explore distant galaxies, our own Milky Way, twinkling stars and mysterious black holes. As Carl Sagan says, "The Earth is just a speck of dust suspended in a ray of sunshine", and this pushes us to contemplate the immensity of the universe around us.

Chapter 2: Beyond Time and Space

Through a philosophical and scientific lens, we examine the concept of time and space. From Einstein's theory of relativity to quantum physics, we discover the implications of a universe where time is relative and space is distorted. As Einstein states, "The past, present, and future are but an illusion, though they are a persistent one," leading us to reflect on the nature of existence and reality itself.

Chapter 3: Depth of Consciousness

We delve into the depths of the human mind and consciousness, exploring subjective experience and perception. From philosophy to neuroscience, we try to understand the mystery of being conscious. As René Descartes reminds us, "I think, therefore I am," underlining the core of our existence and awareness.

Chapter 4: Virtual Reality and Mental Illusions

We investigate the possibilities of virtual reality and mental illusions, exploring how the human mind can be tricked and manipulated. Through examples and case studies, we examine the boundaries of our perception of reality. As Philip K. Dick states, "Reality is that thing that, when you stop believing in it, does not disappear", making us reflect on the changing nature of reality.

Chapter 5: The Search for Extraterrestrial Life

We examine the possibilities of extraterrestrial life and the exploration of exoplanets. Through astronomical and astrobiological research, we explore the conditions necessary for life in the universe and strategies for searching for signs of life beyond Earth. As Carl Sagan reminds us, "There are more stars in the cosmos than grains of sand on all the beaches of the Earth," urging us to keep an open mind to the possibilities of the universe.

Chapter 6: Mysteries of Existence

We address fundamental questions about human existence and the meaning of life. Through a combination of philosophical and spiritual reflections, we explore the mysteries of existence and our place in the universe. As Hermann Hesse reminds us, "Every man is a universe in himself, but society makes him believe that he is only a link in the chain", inviting us to contemplate our uniqueness in the infinity of the universe.

Chapter 7: The Creative Mind

We explore the nature of creativity and innovation through a psychological and philosophical lens. Through historical examples and contemporary studies, we explore the mental processes that underlie creativity and the generation of new ideas. As Pablo Picasso inspires us, "Every child is an artist. The problem is to remain an artist when he grows up", encouraging us to cultivate our inner creativity.

Chapter 8: Knowledge and Ignorance

We examine the concept of knowledge and ignorance, exploring the challenges of our search for truth in the universe. Through philosophical reflections and historical case studies, we explore the limits of our understanding and the nature of human ignorance. As Socrates reminds us, "I know that I know nothing," we recognize the humility necessary in the pursuit of knowledge

Chapter 9: The Beauty of Art

We celebrate the power of art and human creativity to explore the inner and outer universe. Through examples of art, music, literature, and more, we explore how art helps us give meaning to the world around us. As Leonardo da Vinci reminds us, "Beauty lies in the eyes of the beholder", inviting us to contemplate the beauty that surrounds us.

Chapter 10: The Power of Compassion

We explore the role of compassion and love in our journey through the universe and in our understanding of the human mind. Through examples of altruism and solidarity, we explore how compassion can be a transformative force in the world. As the Dalai Lama reminds us, "Our true journey in life is downward, into the heart," encouraging us to spread love and kindness in everything we do.

Conclusion: The Triumph of Exploration

In the book's conclusion, we reflect on the epic journey we have undertaken through the infinity of the universe and the human mind. We realize that despite all our discoveries and experiences, there remains much to explore and understand. With a powerful and inspiring phrase, such as that of Albert Einstein, "Imagination is more important than knowledge", we are reminded that our journey is never truly over, as the search for knowledge and understanding continues to fuel our mind and our spirit.

Introduction: Exploring the Infinity of the Universe and the Human Mind

In the darkness of the cosmos and in the deepest recesses of our minds, lie mysteries that challenge our understanding and fuel our endless curiosity. Welcome to "Over the Horizon", an epic journey that will take us through the infinity of the universe and the human mind, to discover the wonders that await us beyond the boundaries of our understanding.

In this book, we will strive to explore two of humanity's greatest enigmas: the external universe and the internal universe of the human mind. Through a fusion of science, philosophy and imagination, we will delve into the mysteries that lie beyond the stars and thoughts, facing age-old questions and seeking answers that have eluded us for centuries.

The importance of exploring the universe and the human mind cannot be understated. We are inherently curious creatures, eager to understand our place in the infinite scheme of things. The outer universe, with its distant galaxies, unknown planets, and mysterious laws that govern it, offers us a window into the wonders of nature and the complexity of existence. Exploring the universe is an act of discovery and self-knowledge, a way to broaden our horizons and deepen our understanding of the world around us.

But the external universe is only part of the equation. The human mind, with its complexity and capacity for self-reflection, is equally fascinating and mysterious. Through our minds, we can explore worlds of thought and imagination, discovering hidden truths and creating new worlds of possibility. Exploring the human mind is a journey towards the very essence of humanity, an opportunity to understand who we are and what we can become.

Throughout this book, we will examine a wide range of themes and concepts related to the universe and the human mind. We will explore the infinity of outer space, discussing distant galaxies, black holes, dark matter, and other wonders of the universe. We will also examine the mysteries of quantum physics, time and space, and their implications for our understanding of reality.

But we will not limit ourselves to the external universe. We will also explore the inner universe of the human mind, examining the nature of consciousness, perception and knowledge. We will explore the edges of reality and the nature of existence itself, trying to understand what it means to be human in an infinite universe.

To guide us on this journey, we will draw on the words of wisdom of the great thinkers and philosophers of the past and present. As Albert Einstein once said, "Imagination is more important than knowledge, because knowledge is limited, while imagination embraces the whole world, stimulating progress, giving birth to life itself." With these words in mind, we prepare to embark on a journey that will take us beyond the limits of our imagination and discover wonders that far surpass our wildest fantasies.

Prepare yourselves, therefore, to raise your gaze to the starry sky and immerse yourself in the infinity of the universe and the human mind. The voyage of discovery is about to begin, and the wonders that await us are greater than we can imagine. Happy travels, explorers of infinity!

Chapter 1: Journey into Cosmic Infinity

In the deep darkness of infinite space, among the stars that dot the immensity of the cosmos, lies a universe of wonders and mysteries waiting to be discovered. In this first chapter of our journey through "Over the Horizon," we delve into the depths of the outer universe, exploring distant galaxies, remote solar systems, and the mysterious forces that shape the very fabric of reality.

The Galaxies: Universes in Miniature

Imagine looking up into the night sky, looking beyond the familiar stars of the Milky Way, and contemplating the infinity of galaxies stretching out into infinity. Each galaxy is a miniature universe, with billions of stars, planets, and other cosmic structures dancing in interstellar space. Through powerful telescopes and daring space probes, humanity has begun to explore the vastness of the cosmos, discovering galaxies spanning billions of light-years, some dating back to the dawn of time itself.

The Milky Way: Our Spiral Galaxy

At the heart of our local universe lies the Milky Way, our parent spiral galaxy. Known for its iconic beauty and structural complexity, the Milky Way is our cosmic home, one solar system among billions of others. Through astronomical observations and computer simulations, scientists have mapped the many arms of the Milky Way, its clouds of gas and dust, and its star-forming centers.

The Stars: Celestial Furnaces of Light and Life

Stars are the celestial furnaces of the universe, where lighter elements are fused together to form heavier elements, generating light and heat in the process. Each star is unique, with its own mass, temperature and luminosity, and each plays a crucial role in the fabric of the cosmos. From red giants to white dwarfs, from neutron stars to black holes, the universe is populated by an astonishing variety of stars and star systems.

The Planets: Worlds to Explore

In addition to stars, the universe is home to myriad planets, each with its own unique history and characteristics. From the barren rocks of the terrestrial planets to the giant gases of the outer reaches of the solar system, the planets offer endless opportunities for exploration and discovery. Through space missions like Voyager, Cassini, and New Horizons, we have begun to explore the planets in our solar system and beyond, breaking new ground in space exploration.

Black Holes: Portals to Infinity

But among the wonders of the universe also lie its deepest mysteries, including black holes, enigmatic regions of space where gravity is so intense that it bends the very fabric of space-time. Black holes are the eventual fate of massive stars that exhaust their nuclear fuel, collapsing in on themselves and forming a point of gravity so strong that it traps even light. These regions of the universe are among the most mysterious and fascinating, and understanding them is still the subject of intense scientific research and speculation.

Conclusion

In this first chapter of our journey into cosmic infinity, we have only scratched the surface of the wonders and mysteries that await us beyond the celestial horizon. From distant galaxies to enigmatic black holes, the universe offers endless opportunities for exploration and discovery, and our journey just begun is just the beginning of an adventure that will take us beyond the boundaries of our imagination. As Carl Sagan once said, "The Earth is just a speck of dust suspended in a ray of sunshine," yet, in this speck of dust, we find a universe of wonders and mysteries waiting for us to discover. Read on and prepare to be transported to the depths of cosmic infinity.

Chapter 2: Beyond Time and Space

In the infinity of the universe, the concept of time and space takes on a completely different dimension from what we experience in our daily lives. In this chapter, we dive into the depths of relativity and quantum physics, exploring the implications of a universe where time is relative and space is distorted. Through a combination of philosophical reflections and advanced scientific concepts, we will seek to shed light on the mysteries surrounding the nature of time and space.

The Concept of Time: A Persistent Illusion

For centuries, humans have viewed time as a universal constant, flowing inexorably from the past to the present and future. However, with the advent of Albert Einstein's theory of relativity, our concept of time was revolutionized. According to the theory of special relativity, time is relative to speed and gravity, expanding or contracting depending on the circumstances. This concept, expressed by Einstein's equation $E=mc^2$, shook the foundations of classical physics, suggesting that time is not a one-way arrow, but rather a flexible and fluid dimension.

The Theory of General Relativity: Curvature of Space and Time

But it is not only time that is subjected to Einstein's theory of relativity; the space itself is subject to rigorous analysis. According to the theory of general relativity, space and time are intertwined into a single entity called space-time, and the presence of matter and energy curves this entity, creating what we perceive as the force of gravity. This idea of the curvature of space and time paved the way for concepts such as black holes and gravitational waves, suggesting that the reality we perceive may only be a partial representation of the deeper truth.

Quantum Physics: Where Time Hits the Wall of Uncertainty

But if relativity taught us that time is flexible and space is curved, quantum physics has shown us another side of the coin, where time and space dissolve into the fabric of reality. According to the principles of quantum mechanics, time is not a defined quantity, but rather a variable that can only be measured in relation to other quantities, such as energy and position. Furthermore, Heisenberg's uncertainty principle suggests that we can never know exactly the position and velocity of a particle at any given time, creating a fundamental uncertainty in our attempt to understand time and space at the microscopic level.

Philosophical Implications: The Nature of Existence and Reality

Faced with these revolutionary theories, we are faced with fundamental philosophical questions about the nature of existence and reality itself. If time is relative and space is distorted, what does this mean for our perception of reality? If the past, present and future are just a persistent illusion, what is our role in the universe? These are questions that have fascinated philosophers and thinkers for centuries, and understanding them may be central to our understanding of our place in the infinite fabric of space and time.

Conclusion

In this chapter, we have explored the depths of the concept of time and space, from Einstein to quantum physics, reflecting on the philosophical implications of these revolutionary theories. As Albert Einstein once said, "The past, the present, and the future are but an illusion, though they are a persistent one," and yet, in this persistent illusion, we find a universe of wonders and mysteries waiting for us to discover. Read on and prepare to be transported to the depths of cosmic infinity and our understanding of reality itself.

Chapter 3: Depth of Consciousness

In the labyrinth of the human mind, consciousness is the light that guides us through the darkness of the unknown. In this chapter, we will dive into the depths of consciousness, exploring the nature of subjective experience and perception. Through a fusion of psychology, neuroscience and philosophy of mind, we will delve into the depths of our consciousness, trying to shed light on one of humanity's oldest and most fascinating mysteries.

Subjective Experience: The Theater of Consciousness

To fully understand the nature of consciousness, we must begin with subjective experience, the inner world of thoughts, emotions, and sensations that defines us as unique individuals. This theater of consciousness is where our perceptions take shape, where our memories intertwine with our expectations, and where our emotions guide us through life. Through neuroimaging and experimental psychology studies, scientists are beginning to map the neural circuits that underlie subjective experience, paving the way for a new understanding of the mechanisms that shape our consciousness.

The Philosophy of Consciousness: The Enigma of Being

But consciousness is more than just a subjective experience; it is also a philosophical conundrum that has fascinated thinkers for centuries. From Plato to Descartes, philosophers have tried to penetrate the mystery of consciousness, trying to understand what it really means to exist as conscious entities. René Descartes' famous phrase, "I think, therefore I am", invites us to reflect on the nature of our existence, suggesting that consciousness is the very proof of our existence. But what does it really mean to be conscious? This is a question that continues to challenge the limits of our understanding, opening the way to a rich variety of theories and speculations.

Neuroscience of Consciousness: Discovering Neural Circuits

In an effort to unravel the mystery of consciousness, scientists have turned to neuroscience, trying to map the neural circuits that underlie subjective experience. Through neuroimaging studies and computational analyzes of neural data, scientists have identified a network of brain regions involved in consciousness, including the prefrontal brain, limbic system, and thalamus. However, despite advances in understanding the neural mechanisms of consciousness, there still remains a profound mystery about how these brain regions generate the subjective experience that defines our existence.

Awareness and the Self: The Intertwining of Consciousness

But consciousness goes beyond subjective experience; it also includes self-awareness, the ability to reflect on one's experiences and perceive oneself as distinct entities within the universe. This self-awareness is the foundation of our personal identity, informing our choices, our behaviors, and our relationships with the world around us. Through studies of the psychology of mindfulness and meditation, scientists are beginning to better understand the mechanisms underlying self-awareness, paving the way for a greater understanding of the nature of human consciousness.

Conclusion

In this chapter, we have explored the depths of human consciousness, from Plato to Descartes, from neuroscience to the psychology of awareness. Consciousness remains one of humanity's greatest mysteries, an unexplored land of thoughts, emotions and sensations that define us as unique individuals. Read on and prepare to be transported into the depths of the infinite fabric of human consciousness, where mystery and wonder intertwine in an eternal dance.

Chapter 4: Virtual Reality and Mental Illusions

In the vast territory of the human mind, there exist hidden and mysterious corners where the line between what is real and what is illusory becomes blurred. In this chapter, we will dive into the depths of virtual reality and mental illusions, exploring the possibilities of deception and manipulation that the human mind can undergo. Through examples and case studies, we will try to shed light on the boundaries of our perception of reality and the implications this can have for our understanding of the world around us.

Virtual Reality: Intertwined Dreams and Reality

Virtual reality is a world of wonder and possibility, where our wildest fantasies can come to life before our eyes. Through the use of devices such as VR headsets and sensory gloves, we can immerse ourselves in digital worlds that feel so real that it is difficult to distinguish between what is real and what is simulated. However, this fusion of reality and illusion can lead to extraordinary experiences of transformation and personal growth, opening the mind to new possibilities and perspectives that might otherwise remain hidden.

Optical Illusions: When the Eyes Deceive the Mind

But it's not just virtual reality that can deceive our perception of reality; Optical illusions are another example of how the human mind can be easily fooled. Through the use of visual tricks and illusions of perspective, artists and illusionists have created works that challenge our perceptions and force us to question the nature of reality itself. However, these optical illusions also offer us a unique opportunity to explore the limits of our perception and better understand the mechanisms underlying our visual experience.

Cognitive Bias: When the Mind Deceives Itself

But perhaps the greatest deception of all is the one our own minds play on us through cognitive biases. These are mental biases that influence our perception and judgment, pushing us to interpret reality in distorted and irrational ways. Among the most common cognitive biases are confirmation bias, where we tend to seek confirmation for our existing beliefs, and effectiveness overestimation bias, where we overestimate our abilities and skills. These biases can have a significant impact on our decisions and actions, leading us to make irrational choices and ignore evidence contrary to our view of the world.

The Philosophy of Reality: In Search of the Truth

But what does all this mean for our understanding of reality itself? Philip K. Dick's quote, "Reality is that thing that, when you stop believing in it, does not disappear", invites us to question the nature of reality and our perception of it. If reality can be manipulated and distorted, how can we ever know the truth? This is a question that has fascinated philosophers and thinkers for centuries, and its answer could have profound implications for our understanding of the world around us.

Conclusion

In this chapter, we have explored the depths of virtual reality and mental illusions, examining how the human mind can be tricked and manipulated through a variety of means. From immersive virtual reality experiences to optical illusions that challenge our perceptions, we've seen how the line between what's real and what's illusory can be easily blurred. Read on and prepare to be transported into the depths of the infinite complexity of the human mind, where reality and illusion intertwine in an eternal dance.

Chapter 5: The Search for Extraterrestrial Life

In the vast theater of the universe, humanity has always dreamed of discovering life beyond the confines of the Earth. In this chapter, we will dive into the search for extraterrestrial life, examining the possibilities and challenges of exploring exoplanets and searching for signs of life in the universe. Through astronomical and astrobiological research, we will try to shed light on one of humanity's greatest mysteries: are we alone in the universe?

The Significance of Space Exploration: The Journey into the Unknown

Since the dawn of civilization, humanity has looked up to the night sky, wondering about the possibility of life beyond the ends of Earth. However, it is only in the last few centuries that we have begun to embark on a tangible journey into the unknown, exploring space through space probes and advanced telescopes. Space exploration has led us to discover a vast variety of alien worlds, from the icy moons of Jupiter to rocky planets orbiting other stars. These discoveries have pushed us to question the possibility of life outside our solar system, opening the door to a new era of exploration and discovery.

Exoplanets: Potentially Habitable Worlds

One of the greatest achievements of space exploration in recent decades has been the discovery of thousands of exoplanets orbiting other stars in our galaxy. These alien worlds, ranging from Earth-sized to gas giants, offer a unique opportunity to search for extraterrestrial life. Through astronomical research and computer modeling, scientists are identifying exoplanets that may have the conditions necessary to support life as we know it, including the presence of liquid water and moderate temperatures. However, finding signs of life on these worlds remains a monumental challenge, requiring the use of advanced technologies and the analysis of complex data.

The Search for Signs of Life: From Chemistry to Biology

But how can we look for signs of life beyond Earth? This is the question that has pushed scientists to explore new frontiers in chemistry and biology. Through the use of advanced telescopes and molecular analysis tools, astrobiologists are trying to identify molecules and chemical compounds that could be indicative of biological activity on distant exoplanets. This approach, known as biomarker research, could help us detect signs of life even on worlds outside our galaxy, opening the door to a new era of interstellar exploration.

The Philosophical Significance of the Search for Extraterrestrial Life: Reflections on Humanity and the Cosmos

But the search for extraterrestrial life is not just a scientific question; it also has profound philosophical implications on our understanding of ourselves and our place in the universe. Carl Sagan's quote, "There are more stars in the cosmos than grains of sand on all the beaches of the Earth," invites us to keep an open mind to the immensity of the universe and the possibility of life beyond of the Earth. This pushes us to question the meaning of life and consciousness in the universe, opening the door to a rich variety of speculation and reflection on what might lie beyond the boundaries of our understanding.

Conclusion

In this chapter, we have explored the depths of the search for extraterrestrial life, examining the possibilities and challenges of exploring exoplanets and searching for signs of life in the universe. From exoplanet discoveries to the search for biomarkers, we've seen how science is breaking new ground in our understanding of the universe and our place within it. Read on and prepare to be transported into the depths of the infinite fabric of the universe, where life and wonder intertwine in an eternal dance.

Chapter 6: Mysteries of Existence

Deep within the human soul lie mysteries that escape our rational understanding, unanswered questions that push us to explore the boundaries of our being and our place in the universe. In this chapter, we will address fundamental questions about human existence and the meaning of life, exploring through a combination of philosophical and spiritual reflections the mysteries that surround us and our role in this vast cosmic theater.

The Why Question: Searching for Meaning in Human Existence

One of humanity's oldest and most persistent questions is that of why: why do we exist? What is the meaning of our existence on this planet? These questions push us to reflect on our place in the universe and our relationships with the world around us. Through philosophical and spiritual reflections, we seek to shed light on these mysteries, exploring the multiple perspectives that humanity has developed over the millennia.

The Inner Journey: Discovering the Deep Self

But the search for meaning is not just an exploration of the external universe; it is also a journey within ourselves, to discover our true self. Through spiritual practices such as meditation and contemplation, we seek to connect with our deepest selves, beyond the masks and identities that define us in the outside world. This inner journey leads us to face our fears, anxieties and insecurities, paving the way for a greater understanding of who we truly are and our purpose in the universe.

The Reality of Being: Beyond Superficial Appearances

But how much of our existence is really real? This is a question that pushes us to explore the multiple dimensions of reality and to question the fundamental nature of being. Hermann Hesse's quote, "Every man is a universe in himself, but society makes him believe that he is only a link in the chain", invites us to reflect on the depth of our existence and on the superficial appearances that often deceive us. Through the practice of awareness and acceptance, we seek to overcome the illusions of our mind and come into contact with the ultimate reality of being.

The Question of One and All: The Mystery of the Universe

Finally, we are faced with the biggest question of all: what is the bond that unites us to the universe? Are we just single isolated entities or are we part of a larger fabric of interconnection and interdependence? This is a question that pushes us to explore the boundaries of our understanding of the universe and our role in it. Through philosophical and spiritual reflection, we seek to shed light on this ancient mystery, exploring the multiple perspectives that humanity has developed over the centuries.

Conclusion

In this chapter, we explored the mysteries of human existence, addressing fundamental questions about the meaning of life and our place in the universe. From philosophical reflections on the nature of being to spiritual practices aimed at exploring our inner selves, we have sought to shed light on these ancient mysteries that continue to challenge our understanding. Read on and prepare to be transported into the depths of the infinite complexity of human existence, where mystery and wonder intertwine in an eternal dance.

Chapter 7: The Creative Mind

Creativity is a primordial force that lies at the heart of the human experience, a powerful engine that drives us to explore, innovate and imagine. In this chapter, we will dive into the nature of creativity and innovation through a psychological and philosophical lens, exploring the mental processes that underlie the generation of new ideas and the manifestation of art, science, and innovation.

The Source of Creativity: Birth of the Creative Idea

Creativity arises from the meeting of multiple influences and sources of inspiration, a complex process that draws nourishment from human experience and our ability to perceive and interpret the world around us. Through philosophical and psychological reflection, we seek to shed light on this primordial source of creativity, exploring the deep roots of imagination and innovation.

The Psychology of Creativity: The Mental Processes Behind Innovation

But what are the mental processes that underlie creativity? This is a question that has fascinated psychologists and scholars for centuries, pushing us to explore the intricacies of the human mind and its intricate mechanisms. Through case studies and empirical research, we seek to shed light on these processes, examining how the human mind generates and develops new ideas and concepts. From the initial phases of inspiration and intuition to the phase of elaboration and realization, through the search for creative solutions and innovation, we discover the secrets of the creative mind and its infinite possibilities.

The Art of Creativity: Expression and Innovation

But creativity is not just a mental process; it is also an act of expression and innovation that leads to the manifestation of works of art, revolutionary inventions and visionary ideas. Through historical examples and contemporary studies, we explore the many forms that creativity can take, from the works of art of the great masters to the scientific discoveries of innovative geniuses. From Leonardo da Vinci to Steve Jobs, we explore the lives and works of those who have embraced the power of creativity and used it to shape the world around us.

The Philosophy of Creativity: Exploring the Frontiers of the Imagination

Finally, we are faced with fundamental philosophical questions about the nature of creativity and its role in human experience. Pablo Picasso's quote, "Every child is an artist. The problem is remaining an artist when we grow up," invites us to reflect on the intrinsic nature of creativity in humans and the challenges we face in keeping it alive throughout our lives. life. Through philosophical reflection, we explore the multiple perspectives that humanity has developed over millennia on the nature and meaning of creativity, opening the door to a rich variety of speculation and reflection.

Conclusion

In this chapter, we have explored the nature of creativity and innovation through a psychological and philosophical lens, examining the mental processes that underlie the generation of new ideas and the manifestation of art, science, and innovation. From the primordial source of the creative idea to its expression in revolutionary works of art and inventions, we have seen how creativity is a primordial force that lies at the heart of the human experience, a force that drives us to explore, innovate, and imagine. Read on and prepare to be transported into the depths of the infinite complexity of the creative mind, where wonder and inspiration intertwine in an eternal dance.

Chapter 8: Knowledge and Ignorance

In the labyrinth of human knowledge, we are constantly faced with challenges and dilemmas that test our understanding of the world around us. In this chapter, we will examine the concept of knowledge and ignorance, exploring the challenges of our search for truth in the universe. Through philosophical reflections and historical case studies, we will seek to shed light on the limits of our understanding and the nature of human ignorance.

The Challenge of Knowledge: Navigating the Sea of Unknown

The search for knowledge is one of humanity's oldest and most universal challenges. Since the dawn of civilization, humanity has sought to understand the world around it, questioning the laws that govern the universe and the meaning of life itself. However, this quest is often hampered by our own cognitive limitations and the complexity of the world around us. Through philosophical reflection, we seek to shed light on this fundamental challenge, exploring ways in which we can navigate the sea of the unknown and pursue the search for truth.

The Nature of Ignorance: Exploring the Limits of Human Understanding

But what does it really mean to be ignorant? This is a question that pushes us to explore the limits of our human understanding and the nature of ignorance itself. Socrates' quote, "I know that I know nothing," invites us to question our knowledge and awareness of our own limitations. Through philosophical reflection, we seek to shed light on this fundamental question, exploring ways in which we can overcome ignorance and pursue the pursuit of knowledge.

The Illusions of Knowledge: Exploring the Traps of Certainty

But how much of our knowledge is really solid? This is a question that pushes us to explore the many illusions of human knowledge and the traps of certainty. Through historical case studies and contemporary examples, we seek to shed light on these illusions, examining the ways in which our perception of reality can be distorted and deceived by our own beliefs and prejudices. From outdated scientific theories to entrenched cultural beliefs, we explore the ways ignorance can affect our understanding of the world and prevent us from seeing the truth.

The Paradox of Enlightened Ignorance: The Awareness of Our Ignorance

But there is a form of ignorance that can be enlightening: conscious ignorance. This is the recognition of our inability to know everything and the awareness of the limits of our knowledge. Through the practice of intellectual modesty and open-mindedness, we can embrace this form of ignorance and use it as an engine in our search for truth. Awareness of our ignorance drives us to continue to question, explore, and seek to understand the world around us in a deeper and more meaningful way.

Conclusion

In this chapter, we have explored the concept of knowledge and ignorance, examining the challenges of our search for truth in the universe. From the nature of human ignorance to the awareness of our inability to know everything, we have sought to shed light on the limits of our understanding and the nature of knowledge itself. Read on and prepare to be transported into the depths of the infinite complexity of the search for truth, where wonder and uncertainty intertwine in an eternal dance.

Chapter 9: The Beauty of Art

Art is a beacon that illuminates the path of humanity through the centuries, a manifestation of the creative power of the human soul that pushes us to explore the inner and outer universe. In this chapter, we will celebrate the power of art and human creativity to explore the world around us, through examples of art, music, literature, and more. Through the beauty of art, we discover how we can give meaning to the world and our existence.

The Expression of the Soul: Art as a Vehicle of Emotions and Thoughts

Art is a form of human expression that allows us to communicate emotions, thoughts and ideas in a unique and powerful way. Through painting, sculpture, music, literature and other forms of artistic expression, we give voice to our deepest experiences and our most intimate visions of the world. Art allows us to explore the inner world of the human soul, revealing the beauty and complexity of our emotions and thoughts.

The Wonder of Creativity: Art as an Act of Innovation and Discovery

But art is not just a means of expression; it is also an act of creativity and innovation that allows us to explore new territories of the mind and soul. Through art, we discover new perspectives on the world and our experiences, opening the door to greater understanding and awareness of ourselves and the world around us. Art invites us to see the world through the eyes of the artist, to embrace the beauty and wonder that surrounds us every day.

Art as a Reflection of Nature: Exploring the Beauty of the Natural World

One of the greatest sources of inspiration for artists of all time has been the beauty of the natural world. Through landscape paintings, animal sculptures, and nature-inspired musical compositions, art allows us to explore the beauty and grandeur of the world around us. From towering mountains to endless oceans, from the Northern Lights to the delicate blooming flower, art invites us to contemplate the beauty and perfection of nature and recognize our place within it.

Beauty as a Portal to Transcendence: Art as a Spiritual Experience

But art is also a portal to transcendence, a means through which we can connect with something greater than ourselves. Through the beauty of art, we discover the infinite and the eternal, the divine that resides within us and around us. Leonardo da Vinci's quote, "Beauty lies in the eye of the beholder," invites us to contemplate beauty as a manifestation of the divine that resides within us and around us. Art allows us to experience unity and connection with the universe, opening the door to a profound spiritual and transcendental experience.

Conclusion

In this chapter, we explored the beauty of art and human creativity, celebrating the power of art to give meaning to the world and our existence. Through examples of art, music, literature and more, we have seen how art allows us to express our deepest emotions and thoughts, to explore new territories of the mind and soul, and to connect with something greater than ourselves themselves. Read on and prepare to be transported to the depths of the infinite beauty of art, where wonder and inspiration intertwine in an eternal dance.

Chapter 10: The Power of Compassion

In the fabric of the universe and the human mind, compassion shines like a bright star, a transforming force that guides our journey through the depths of existence. In this chapter, we will explore the role of compassion and love in our journey through the universe and in our understanding of the human mind. Through examples of altruism and solidarity, we will discover how compassion can be a powerful and transformative force in the world.

The Nature of Compassion: The Expression of Universal Love

Compassion is the ability to understand and respond to the suffering of others with kindness and empathy. It is a reflection of the universal love that resides in the heart of every human being, a force that connects us with others and the world around us.

Through the practice of compassion, we can open our hearts to others and spread love and kindness wherever we go.

Compassion as a Transforming Force: The Power of Altruism and Solidarity

But compassion is more than just a feeling; it is also a transformative force that can change the world. Through examples of altruism and solidarity, we see how compassion can inspire actions of kindness and generosity that have a lasting impact on the lives of others. From small, everyday acts of kindness to large-scale social solidarity initiatives, compassion shows us the power of love to create a better world for all.

Compassion in Spiritual Practice: The Heart as the Center of Wisdom

In spiritual traditions around the world, compassion is seen as a fundamental aspect of spiritual practice. Through meditation and contemplation, we can cultivate compassion in our hearts and spread love and kindness throughout the world. The Dalai Lama's quote, "Our true journey in life is downward, into the heart," invites us to explore the power of compassion in transforming ourselves and the world around us.

Compassion as a Path to Wisdom:
Learning from the Experience of Others

But compassion is also a path to wisdom, a way to learn from others and their experience. Through empathetic listening and sharing our stories, we can learn to better understand the challenges and suffering of others and respond with compassion and kindness. In this way, compassion becomes a bridge that unites us with others and allows us to grow together in love and mutual understanding.

Conclusion

In this chapter, we have explored the power of compassion in our journey through the universe and in our understanding of the human mind. Through examples of altruism and solidarity, we have seen how compassion can be a transformative force in the world, inspiring actions of kindness and generosity that have a lasting impact on the lives of others. Read on and prepare to be transported to the depths of the infinite beauty of compassion, where love and kindness intertwine in an eternal dance.

Conclusion: The Triumph of Exploration

Throughout our journey through the pages of "Over the Horizon: A Journey into the Infinity of the Universe and Mind", we have immersed ourselves in the depths of human existence and the universe around us. Through philosophical reflections, scientific explorations and spiritual speculations, we have sought to shed light on the deepest mysteries of life and existence, exploring the infinity of the universe and the human mind.

Beginning our journey with a glimpse of the wonders of the outer universe in "Journey to Cosmic Infinity," we contemplated the vastness of distant galaxies and enigmatic black holes, reflecting on our place in the cosmos. We then moved "Beyond Time and Space", exploring the implications of the theories of relativity and quantum physics, which showed us how time and space are relative and fluid concepts, distorting our perception of reality.

In the chapter "Depths of Consciousness", we delved into the depths of the human mind, exploring the nature of subjective experience and consciousness itself. Through a combination of psychology, neuroscience, and philosophy of mind, we have sought to understand the intricacies of our consciousness and its deepest mysteries.

Next, in the chapter "Virtual Reality and Mental Illusions", we investigated the possibilities of virtual reality and mental illusions, exploring how the human mind can be deceived and manipulated. Through examples and case studies, we explored the boundaries of our perception of reality, revealing how fragile and subject to distortion it is.

Continuing our journey, we explored the possibilities of extraterrestrial life and the exploration of exoplanets in the chapter "In Search of Extraterrestrial Life", opening our minds to the possibility of life beyond our planet Earth. Through astronomical and astrobiological research, we have explored the conditions necessary for life in the universe and strategies for searching for signs of life beyond our Earth.

In the next chapter, "Mysteries of Existence", we grappled with the fundamental questions about human existence and the meaning of life. Through a combination of philosophical and spiritual reflections, we have sought to shed light on the mysteries of existence and our place in the universe, recognizing that many questions remain unanswered, but that it is in seeking to answer them that we find meaning.

In The Creative Mind, we explored the nature of creativity and innovation through a psychological and philosophical lens, examining the mental processes that underlie the generation of new ideas and the manifestation of art, science and innovation. We recognized creativity as a primordial force that lies at the heart of the human experience, a force that drives us to explore, innovate and imagine.

Next, in the chapter "Knowledge and Ignorance," we explored the concept of knowledge and ignorance, examining the challenges of our search for truth in the universe. Through philosophical reflections and historical case studies, we have shed light on the limits of our understanding and the nature of human ignorance, recognizing the humility required in acknowledging our limited understanding of the world.

In the chapter "The Beauty of Art", we celebrated the power of art and human creativity in exploring the inner and outer universe. Through examples of art, music, literature and more, we explored how art helps us give meaning to the world around us, recognizing beauty as a bridge that unites humans and the divine.

Finally, in the chapter "The Power of Compassion", we examined the role of compassion and love in our journey through the universe and in our understanding of the mind

Human. Through examples of altruism and solidarity, we have seen how compassion can be a transformative force in the world, inspiring actions of kindness and generosity that have a lasting impact on the lives of others.

In each chapter, we have sought to shed light on the mysteries of existence and the human mind, exploring the depths of the universe and our own consciousness. We have been transported through the wonders of science, philosophy and spirituality, searching for answers and meaning in the vast ocean of existence.

As we conclude this journey, we can reflect on our shared discoveries and experiences, recognizing that the journey of knowledge is infinite and that beauty and wisdom lie in the deepest questions that drive us to explore, seek and grow. We can take inspiration from the words of Albert Einstein: "The most beautiful and profound journey is the one we undertake in the heart".

May this journey light a flame in your mind and heart, carrying you forward with curiosity, compassion, and gratitude for the infinite beauty and mystery of the universe and the human mind. And may you continue to explore, discover and celebrate the miracle of existence with every step you take.

Over the horizon
A Journey into Infinity
of the Universe and the Mind
Eng. Das Warhe

Eng. Das Warhe, 2024
All rights reserved.
Reproduction is permitted for educational and non-commercial purposes only, provided that the source is cited
April 2024

VAS Publisher

www.ingramcontent.com/pod-product-compliance
Lightning Source LLC
Chambersburg PA
CBHW070215230526
45471CB00002B/955